The Science of

Birdnesting

By

Major H. T. GOSNELL

Member of the British Ornithologists' Union.
Member of the Jourdain Society, formerly the
British Oological Society.

INTRODUCTION

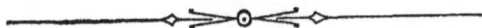

Since this book was started, two events have occurred which have had a profound effect on bird-life, chiefly in Southern England.

1. The great frosts of winters 1939/40 and early 1947.

2. The Second World War.

Dartford Warblers and Stonechats were obliterated by the former in Hampshire, Sussex and Surrey and have been prevented from re-establishing themselves from the West by the almost total destruction of our beautiful Commons by Armoured Fighting Vehicles.

I returned from six years War Service to find these changes, which include the felling of nearly all pine trees of any size.

Where the words ' never ' and ' always,' are used, it must be understood that these are the result of my own observation and must not be regarded as infallible. Also I have not quoted knowingly the opinions of other ornithologists except where recognition has been given.

I wish to record my gratitude to my wife for all her assistance, and to my friends for the loan of their photographs.

March, 1947.   H.T.G.

FOREWORD

I have frequently been asked by boys interested in the study of our breeding birds, " how to find the nest of such-and-such a bird.' so that I have attempted to try and answer them in this little book.

The list of species is purposely incomplete: common hedge-row birds have been omitted, also rare ones, and birds that breed in colonies or on our sea cliffs, for obvious reasons; and in the interests of economy, in order that the price of this book may be within the reach of boys' pockets. Lengthy descriptions of the birds themselves have been avoided, and only brief aids to identification in the field given, as these can be studied more fully in numerous excellent works such as ' The British Bird Book ' by Kirkman and Jourdain, to mention one which stands out as unsurpassed in value.

The subject as treated herein is intended as a guide to the technique of nest-finding, and to be an encouragement to the individual to work out a system for himself.

H. T. GOSNELL,
THE BOREEN,
HEADLEY DOWN,
BORDON, HANTS.

The Science of Birdnesting

THERE are two main reasons for collecting the eggs of wild birds. First, to show the variation in colouring, etc., of each species. Second, to use the egg of the parent bird as a means of tracing it throughout its life.

This range of variation in some species is extraordinary, and it is quite impossible to show a fair sample of their eggs by only a few specimens. Others, like the birds laying white eggs, vary not at all, and one set or so is all that is required. Amongst the eggs of those birds that have but little variation, the same applies, only a single clutch should be taken, and search for really fine specimens concentrated on.

This is one of the fascinations of egg-collecting, the beginner has the same chance of finding a unique set of some common bird, as the man who has been at it all his life.

One can look through scores of Blackbirds' and Thrushes' nests each season without finding anything worth a place in the cabinet. But who knows, anyone might be as fortunate as the late Mr. Streeter, ot Petworth, Sussex, who found near his home a Song-Thrush which laid pure white eggs with red spots. Several sets of these wonderful eggs are in different collections and doubtless there would have been more, if some vandal with a gun had not shot the bird.

Every season I examine dozens of Robins' nests, when following up the local Robin-Cuckoos. Yet only once in my life-time have I found a Robin laying pure white eggs.

Again, take Willow-Warblers, probably the commonest migrant in the country. Of all the nests I have found, only two have held 8 eggs, yet quite 30 per cent. of nests in this district contain 7 eggs, and I have never seen a pure white unmarked clutch in situ.

Now turn to the Nightingale, how often can any variation be found in their eggs? All are dull olive green. One of my friends found 14 nests of this bird in one season in Sussex, all as like as peas in a pod. Yet the late Col. C. Smeed, another Sussex man, found a beautiful set with pale blue eggs all unmarked.

One day on the South Downs I found the nest of a Skylark with four wonderful eggs, (mauve and brown blotches on a pale ground), having no resemblance to the well-known greeny eggs of this species. So strange were they, that when I showed them to a very well-known collector, he said "Those are not Skylark's eggs—they are some foreign bird"!

I could go on and on, but the foregoing should show the principles on which to work.

The museums can get their material only from private collectors, and it is up to us to see that they are offered the chance of obtaining really scientific collections of local birds.

How many museums contain such collections? I know of none outside South Kensington.

The second reason for egg-collecting, that of tracing the parent bird has been much neglected in the past, and its use opens up a fruitful field of inquiry for the future. When a bird is discovered which lays an egg in some way different from others of its kind,

by which it can be unfailingly recognised, it should be the object of the most careful observation, for that bird is as good as labelled, or ringed.

Articles have recently appeared in the scientific Press giving the writers' opinion of the ages reached by certain common garden birds. My own view is that his opinion is very far from the truth. Here the egg-collector can help. I know a Cuckoo that has been laying here for thirteen seasons, and of a Crow for fourteen. Both these birds were traced by their eggs, which varied sufficiently from the normal for the birds to be recognised throughout their laying life.

What advice then should be given to the young naturalist who is keen on bird-nesting? If he is only interested in finding nests, and regards the eggs merely as trophies of the chase, then I strongly advise taking only one egg from each nest.

These can be cheaply housed in a small cabinet made with partitions of wood which can be altered to suit requirements, or even in home-made boxes.

But if having read what has gone before, the collector feels that a scientific study is more in his line, then he should concentrate on searching for variety, keeping down type eggs to the minimum, and collecting in clutches, blowing with one hole, and filling up a data card for each clutch. He will not amass a large number of eggs by this means (so much the better), but what he does take will be worth having, and will add something to our knowledge.

Many people are under the impression that if all the eggs are taken from a nest, the parent birds are deprived of their young for the season. This, of course, is quite erroneous, as birds, (with very few exceptions) all replace lost eggs, nests or young, in a very few days; ten days as a rule for a small bird such as a lark or pipit, and proportionately longer for larger species. If this were not the case, birds would long

ago have ceased to exist, so many enemies have they to to contend with in nature, all after their eggs or young. Here are a few—Crows, Magpies, Jays, Cuckoos, Mice, Rats, Voles, Snakes, Cats, Hedgehogs and many others. Some of the above spend the long spring and summer days in a ceaseless search for eggs, and are far better bird nesters than humans.

So when well-meaning friends say " Surely you do not take all the eggs in one nest?" you can tell them that they will easily and rapidly be replaced; also if one or two are abstracted from a nest, the bird will hatch one or two less young in that brood. But if all are taken the repeat lay should contain the same number, or even more eggs than in the lost nest.

It is significant that the natural enemies of birds always empty the nest.

During bird-nesting, useful information will be gathered on some of the important aspects of bird-life, at present only partially investigated. This is where the note-book comes in, extracts from which should be written up more fully in the Day-book on returning home. Here are some examples:

Which parent(s) take part in the nest building?

How long does the nest take to build?

Interval between deposition of eggs, and time of day laid.

When does incubation start?

Share of the sexes in incubation, and does the cock feed the hen on the nest?

Incubation period in days and hours.

Fledgling period.

Share of the parents in feeding and brooding young in nest.

Duties of parents when young first leave nest.

Interval between broods.

Time taken in replacement of clutch when nest has been destroyed, etc., etc.

Here is a list of some birds which are likely to be nesting during the Easter holidays. The approximate date when the clutch should be complete is also given.

Period 25th March—15th April.

Hooded Crow	15th April.	
Carrion Crow	10th April.	
Rook	24th March.	
Woodlark	30th March.	
Mistle Thrush	8th April.	Often last week of March.
Stonechat	11th April.	
Robin	6th April.	
Dipper	1st April.	
Long-eared Owl	25th March.	Variable.
Tawny Owl	30th March.	
Lapwing	3rd April.	
Snipe	10th April.	Variable.
Woodcock	25th March.	Variable.
Moorhen	10th April.	Variable.

Period 16th April-5th May.

Jackdaw	27th April.	
Magpie	25th April.	
Jay	5th May.	Mostly a week later.
Greenfinch	25th April.	
Linnet	20th April.	
Skylark	1st May.	Variable.
Meadow-Pipit	25th April.	
Grey Wagtail	25th April.	
Pied Wagtail	25th April.	
Tree Creeper	20th April.	

Nuthatch	30th April.	Variable, sometimes March.
Great Tit	30th April.	Mostly later.
Marsh Tit	5th May.	Mostly a week later.
Long-tailed Tit	10th April.	Variable.
Goldcrest	5th May.	
Chiff-Chaff	5th May.	Mostly a week later.
Dartford Warbler	25th April.	Variable.
Wren	25th April.	Variable.
Kingfisher	20th April.	Variable.
Little Owl	1st May.	
Barn Owl	1st May.	
Common Buzzard	25th April.	
Mallard	20th April.	
Shoveler	25th April.	
Little Grebe	25th April.	Variable.
Wood Pigeon	15th April.	
Stock Dove	1st May.	
Stone Curlew	1st May.	Variable.
Ring Plover	20th April.	
Redshank	30th April.	
Curlew	20th April.	
Coot	20th April.	Variable.

It will be seen that the period 25th March-5th May has been divided roughly into two parts of about three weeks each.

The approximate date for full clutches is intended only to be a guide, as it is quite impossible to be more accurate owing to:

(a) Variation in the weather year by year.

(b) Different localities such as South, West, and North of England, etc.

(c) Variation of nesting time in the same species owing to the idiosyncracy of individual birds. The half period is intended to be the main guide and the

THE AUTHOR IN DARTFORD WARBLER COUNTRY.
Photo by Major H. M. Heyder, M.C.

NEST OF WOOD-WARBLER.

Photo by Rev. C. J. Pring.

NEST OF TREE-PIPIT.

Photo by Rev. C. J. Pring.

reader must work out the date according
to district and season; for example,
Common Snipe is placed in the first
half of the period, and the 10th April is
given as a good average date for eggs,
yet in the South of England some Snipe
in a dry season will be hatched by that
date, or in the North Country, not sitting
by then. But taking the average of
several years the 15th April is a good
date.

General Hints.

An egg should never be taken unless something
can be learned from it.

It pays to walk ' off ' a track—it is surprising
the number of nests found within a pace or two of
a path.

A light shooting-stick can be a great boon, and
save many a weary stand.

An efficient pair of prismatic field-glasses are
essential for correct identification in the field, also
useful for watching back certain ground-nesting
species at long range.

The new short climbing-irons are a great improve-
ment on the old long ones, and can easily be carried
in a small haversack.

Note-books should always be carried in the pocket
and details of nest, site, etc., written down **on the spot.**

Data tickets should go with every egg or clutch,
as the case may be, and these should be carefully
and accurately filled in. A number on the ticket
should correspond with a number on the eggs. Data
tickets can be obtained ready printed from any of
the recognised firms who deal in natural history.
The price is modest. A standard example is here
shown.   *b*

OOLOGICAL COLLECTION OF

Date........................

No. of Eggs............ Species ⎫ ....

Set Mark................. ⎭ ....

Incubation........................... Identity........................

Locality ....

Nest ....

....

....

....

Under the heading 'Identity' the space following should never be filled by the words ' certain ' or 'sure,' as this in practice means nothing. We are all liable to make mistakes, and the best ornithologists are no exception. We may honestly believe at the time that the eggs in question belong to the Marsh-Tit, let us say, and later we write ' Identity certain ' on the data ticket. But what is wanted is **how** did we identify the said eggs as Marsh-Tits, as this bird is for practical purposes indistinguishable to the eye from the Willow Tit, and only by their voices can the two be recognised. What we should have written is —Identity—' by call notes,' and on the back of the card perhaps, ' chicabee-bee-bee ' or something like it. This would prove to all acquainted with the species that our identification was correct. Such detail would be unnecessary with birds that are in themselves distinctive and not to be confused with any other, such as the Goldfinch; or it may be that the eggs alone would be sufficient evidence, as in the case of the Sparrow-Hawk; in these examples the word ' bird ' in the first, and ' eggs ' in the second, would be scientifically correct.

Some species, of course, can be recognised by their nests alone, such as the Song-Thrush, Nightingale, etc., then the word ' nest ' would suffice.

The Preservation of Eggs.

Eggs should be blown reasonably soon aft r taking, as if left for days without turning, the yolk is liable to congeal and stick to the shell. This, in the case of small eggs makes the task of cleaning very difficult if not impossible.

The egg should be held by the middle finger and thumb at the apices (top and bottom), and a small hole made with the drill in the middle of the side which is not required to lie uppermost in the cabinet. The hole should be as small as is consistent with the complete removal of the contents. The hole should face downwards, and air pressure directed on the outside of the hole, when air is forced in, and the contents gradually expelled. In some eggs which are of a glutenous nature, such as Swifts' and Nightjars' eggs, difficulty may be experienced in getting the ' white ' to start to flow. This can be overcome by taking a small sip of water in the mouth and blowing it into the egg. When the egg is empty this process should also be used to wash it clean, or a little water blown into the egg, and then well shaken (with a finger over the hole) before blowing out.

In the event of the egg being a few days incubated, the membrane of fine blood vessels sometimes has an aggravating habit of remaining behind, and refusing to move, in spite of the application of much water and air. This skin must be removed, otherwise it will attract insect life, or in the case of white eggs, spoil their appearance. The best way I have found to deal with this trouble, is to fill the egg to overflowing with cold water, and then stand it, hole uppermost, on a piece of cotton wool. Return in an hour or so, or even the next day, and blow out the water, when the

offending membrane will come away at once. The
eggs now being empty must be put to drain. An old
table will be found handy for this purpose. A handkerchief ,also old, should be stretched by means of several
drawing pins round the edges, on to the table. One
by one after blowing is finished the eggs should be
placed hole downwards on the smooth handkerchief,
which acting as a sponge draws out any remaining
moisture. If table and handkerchief are not forthcoming, owing to parental disapproval or other means,
a good substitute may be made with a box lined with
a sheet of blotting-paper shaped to the inside bottom
of the box, and fixed with a little paste to lie flat and
smooth. This will act as a drain-board nearly as well
as the handkerchief, its only drawback being that the
slightest vibration in the room may cause some of
the eggs to roll. It must here be stated that the box
should have a lid of some sort, as every endeavour
must be made to keep all coloured eggs away from
light as much as possible. All eggs fade, more or less,
once they are blown, and light accentuates this. Once
the eggs are thoroughly dry, it only remains to write
an identification number on them, corresponding with
that on the data ticket.

Incubated Eggs. Large and Medium Size.

If incubation has advanced too far for blowing in
the ordinary way, even with an enlarged hole, the aid
of chemicals can be sought. It must be emphasised
that the use of these alkaloids is apt to weaken the
shell, and should be avoided if possible.

The water method.

All eggs can be blown in time, given enough
patience and endurance, by filling up with water and
leaving to rot. This process can be assisted by a
daily shaking. The use of an attic, summer-house or
similar isolated spot is indicated, and this method

should not be attempted by those with sensitive noses!

Not having time or inclination for the ' water method ' the alternative is an egg solvent. This is a substance that can be purchased cheaply from any dealer, but it has the disadvantage of being very unpleasant to handle, and must be used with care. If any is spilled it will burn whatever it touches, hands, clothes, furniture, carpets, etc.

A glass suction pipe, which is included in all egg-blowing outfits, must now be used. A small quantity, a teaspoonful, for an egg the size of a Sparrow-Hawk, must be sucked into the glass tube, and prevented from running out by placing the thumb over the top. The point is inserted into the egg with one quick movement to avoid drips, and the contents gently blown into the egg. If any of it falls on the outside of the shell it should be wiped off at once with cotton-wool. The process should be repeated if the egg will hold another dose, but not overfilled, or the appearance of the shell may be spoilt. The chemical safely in the egg, it must be put away on a bed of cotton-wool, so that it cannot roll about, (hole uppermost of course) for about two days. Then it should be held upside down and the mouth filled with water to blow gently into the egg. This will remove the somewhat unpleasant contents. If anything remains, not properly dissolved, the whole process must be repeated. But unless the egg is very large, one application is usually sufficient. The egg must now be well rinsed, half-filled with clean water, shaken and rinsed again, until all traces of the chemical are removed, then it should be placed to drain and dry.

Small eggs.

The problem of the preservation of incubated eggs of the small birds such as Warblers, Pipits,

Wrens, etc., is a very difficult matter, and I have never heard of any really satisfactory method.

It therefore behoves all ornithologists to refrain from taking any eggs of these birds when in an advanced state of incubation

What is an advanced state of incubation?

Anything more than a week ' sat ' should not be taken. It may be possible sometimes with great skill or luck to save an egg a day or so more incubated, but the eggs are useless as specimens as a rule. The shells of these little eggs such as Willow-Warblers, etc., become so brittle that when in the process of blowing them, air-pressure is applied to the drill-hole, they usually burst.

It will then be asked: ' How am I to know when eggs are highly incubated?' If the egg when taken from the nest and held to the sun, shows any transparency, then it is blowable. In the case of a fresh egg, the whole yolk is clearly visible, this becomes thicker as incubation progresses until finally no light can be seen, the egg appearing as solid as a stone. The texture of the shell also alters and the eggs take on a leaden hue and a high shine; both these symptoms are unpropitious from the collector's point of view.

Then there is the ' rolling ' test, which I have found very reliable. Holding the egg in the middle of the outstretched palm of the hand, it should be allowed to roll gently from side to side. If it moves in a series of jerks,, it will be heavily incubated. But if it rolls smoothly and evenly from side to side it should be fresh.

The ' water test.' If water is available near at hand, place the egg gently in (if deep with hand beneath. If the egg sinks to the bottom it is fresh. If it floats mid-way between the surface and the bottom it is nearly fresh. If it floats on the surface with difficulty—a few days incubated. But if it floats

buoyantly like a cork it should not be taken. In actual practice the last sentence is not always accurate, as some eggs can be blown although they float, but for practical purposes it is a useful guide.

If in spite of all care taken in selection, it is found on drilling the hole that no liquid comes out, then difficulty may be expected. In nine cases out of ten it will pay to give up the unequal contest, and throw the egg away, though this is a painful decision if dealing with a fine type or a new specimen. The shell would probably collapse on application of the drill, so in any case a trial could do no more than waste a little time.

But in the tenth case, perhaps a Cuckoo's egg which has survived the drilling, how can it be preserved?

The answer is an ants' nest in the garden, usually found underneath stones. The egg should be placed thereon, hole downwards but not resting on the ground. For if the hole is blocked up by the ground it lies on, the ants are unable to enter the egg, and will soon bite a hole or holes for themselves which would ruin the shell. The egg should be lightly covered with a box lid, to prevent accidents, and visited occasionally to view progress. Time of cleaning will depend on the number and voracity of the ants. I have had a Cuckoo's egg cleaned for me by this method in four or five days, and it was a tiny ants' nest of the black variety. It is as well to mention here that it is not advisable to wait for the inside to be picked clean, or there is a risk of damage to the shell by hungry ants searching in vain for succulent morsels. The last scraps can always be washed out, by filling up the eggs with water and leaving to soak, afterwards washing until quite clean.

Finding nests.

There are three ways of discovering the whereabouts of a nest:

1. By use of the eyes.

Example: The nest of a magpie.

Is obvious to all, before the leaf is fully out, like a clothes-basket at the top of a tree, or high straggly hedge.

2. By accidental flushing of the sitting bird.

Example: The nest of a Willow-Warbler.

Nearly everyone who has wandered in the woods during May or June, has at some time or other seen a little greeny-brown bird departing hurriedly from under their very feet, and on looking carefully at the spot, finds the nest most cunningly concealed.

3. By the use of science.

That is to say by making use of knowledge of the song—habits—call and alarm notes, first to locate the birds themselves, and secondly to watch the sitting parent (usually the hen) return to its nest. It is with the last method that this book is chiefly concerned, and I have tried to explain in the case of each bird, in alphabetical order—

General appearance.
Chief differences of closely allied species.
Song.
Call and alarm notes.
Habitat.
Site of nests.
Number of eggs laid.
Best date to find eggs.
General field notes.

FIELD NOTES.

The foremost need for the discovery of the whereabouts of birds, is the ability to recognise their song and calls. It is unfortunately not possible to give really accurate renderings of these in terms of human

speech. Attempts have been made from time to time to do so, without much success. The Author, therefore, while giving descriptions briefly in words, relies chiefly on the comparison of one bird's song and calls with that of another, possibly better known. For example:

The alarm notes usually written down as ' chip-chip-chip ' could apply equally to the Crossbill, Corn-Bunting and Tree-Pipit, but in the field there is a vast difference between the tone and volume of all three

An ornithologist to be successful must know to which birds belong the various songs, calls, etc., he hears as he wanders about. This may appear very difficult to the beginner, especially in the Spring when the countryside resounds with bird music. In practice, however, provided the student has normal hearing, it is not too onerous. Most of the sounds will be found to come from quite common birds, and these sounds must be learnt and tabulated in the brain, which may be likened to a wireless receiving set, so that it becomes more and more sensitive; soon it will be found that a strange note, being unexpected, will compel attention, and then more than half the battle is won. It should be made a rule to try and identify the owner of any notes which are unfamiliar. As this will entail much time spent in creeping from tree to tree (only to find that the singer has moved to another and thicker tree!), the watcher is not likely to be caught out more than twice by the same call.

THE CHIFF-CHAFF.
(Phylloscopus c. collybita)

Field characteristics.

Very like Willow-Warbler, but more restless and elusive, and not nearly so numerous. At close range its black legs make identification certain.

Allied species.

See under Willow-Warbler.

Song.

Very monotonous, a repetition of ' chiff chaff ' or
' chiff chiff chaff.'

Call and alarm notes.

A soft ' wheet ' repeated quickly over and over
again. This becomes much louder after the young
are hatched. The nearest call that I can think of
that might be mistaken for it is the ' wheet ' note of
the Chaffinch, but that is much slower in tempo and
of rather different quality. I found this bird swarming
in Brittany in the Spring of 1940, and my ears got
thoroughly attuned to its call. I next met it in Sussex
in June, 1945, when in our concentration area, and
could not detect any difference between them.

Eggs.

Usually 6, but 5 or 7 occur. Clutches are
normally ready about mid-May. Sometimes they are
sitting late in April in an early season like 1945. I
think that this bird is more often double-brooded
than is generally realised.

Nest finding.

As a rule not too difficult a nest to discover,
owing to its love of building along the edges of tracks,
paths and roads. The nest, unlike the Willow-Warbler
and Wood-Warbler is nearly always off the ground in
some bramble, butcher's broom, etc. Also when the
hen is off the nest she calls more consistently than
the Willow-Warbler, so that she should be picked up,
or at any rate her approximate position found.

This bird I find, is much shyer than the Willow-
Warbler, and keeps higher in the trees before returning to the nest. It does not pay to stand too
close. The cock sings at times a long way from the
nest.

If destroyed, the nest and eggs are replaced in 10 days. The cock does not 'incubate.

◇➤○≺◇

THE CIRL BUNTING.

(Emberiza c. cirlus)

Field characteristics.

A bird of the high tree-tops which is seldom seen close to, except in the breeding season, when the black bib of the cock is very noticeable. The hen is sparrow-like, but on the ground it has a longer hop, and seems to be sitting back on its tail with a somewhat unbalanced effect. Its legs are paler than a House-Sparrow's, and its flight is noticeably more undulating than a Yellow-Hammer's.

Allied species.

The Yellow-Hammer is the only likely source of confusion. The hens are very similar, but in both sexes the rumps of Cirls are olive, and Yellow-Hammers are brown.

Song.

Many think that the song of the Cirl is like a Yellow-Hammer's. I lived for eight years in Sussex, where this bird is common, and as I had a nest every year in my garden, its notes were ever in my ears; I cannot detect the slightest resemblance. The song of the Greenfinch has a passing likeness, but once the real Cirl tunes up there should be no confusion at all. It is shriller, louder, and of a different quality, and sounds more cheerful. The Lesser Whitethroat's song does resemble that of the Cirl, but is a much slower and deeper note.

The Cirl will sing on any day throughout the year if the weather is kind. After the breeding season its song loses some of its power.

My nearest rendering of the song is: ' Sis-sis-sis ' very quickly about ten times repeated.

Call and alarm notes.

A sharp 'zit,' not unlike the Yellow-Hammer's but shriller.

Best dates for eggs in a normal season:

1st Brood — 18th May.

2nd brood — 25th June.

3rd brood — 1st August.

Clutch 3 or 4, sometimes 5 in second brood. The eggs are often confused with Yellow-Hammer's by those who do not know them well, but they should not be, if examined carefully. Many have a greenish-back-ground. The black lines and blotches are thicker and more intense. And, most important, they **never** have those mauvish lines and smears so often seen on Yellow-Hammer's eggs. Fortunately there is an infallible method of distinguishing between the nests of the two species. The nest of the Cirl-Bunting ALWAYS has a considerable amount of moss **under** the cup. The Yellow-Hammer NEVER. R. Carlyon-Britton pointed this fact out to me many years ago.

To illustrate how even an expert on this bird can be mistaken, I recount the following story:—

One day I was looking over some eggs, unblown, and recently taken by him. He said: "What do you think of that fine set of Cirl's eggs?" I replied: "I see no Cirls, but a grand clutch of Yellow-Hammer's.": He went off to settle the point by getting the nest. It was an undoubted Yellow-Hammer's. I tell this only to emphasise the extreme care needed in the proper identification of eggs in closely allied species.

If destroyed the nest and eggs of the Cirl-Bunting are replaced in 10 days with unfailing regularity.

Nest finding.

The favourite singing post of the cock I should expect to be close to the nest. But unfortunately for the seeker, he has several. Also he is apt to sing for a longish time at any of them. I fear that there is no getting away from the fact that if a Cirl's nest is wanted it must be searched for, and thoroughly too.

The favourite site in Sussex is a thin straggly hedge, often so transparent that to one unacquainted with this bird, it would not seem worth looking at.

The hen is a close sitter, and will allow you to beat over her without always leaving the nest.

Elm hedges are good too, and so are the suckers that grow from the base of these trees. Haystacks are sometimes used, after the manner of a Pied-Wagtail. In my garden the nest occurred in Apple, Macrocarpa, Hawthorn hedge, and one in September in a thin bramble.

During building, the cock is much tamer, and may be seen furtively following his mate, who has a piece of hay or moss in her bill. They then pay little heed to the observer, provided that he keeps reasonably still.

One April, two fine cocks fell at my feet locked in combat, but both flew off unharmed when I stooped to pick them up. This bird comes near to nesting in colonies, several breeding pairs being found over a small area. These colonies may be separated by a mile or two ,with perhaps an odd pair here and there acting as links.

Now and then a nest may be found with amazing ease. I once went to a friend's house some way from my home in a spot that was new to me. As I rang the bell I heard a Cirl singing from the top of a lofty elm by the door. I walked across the drive and peeped into the hedge opposite. There was the nest ready for eggs.

I once found a nest built into a handful of hay that had been blown off a cart and stuck on the top of a low hedge. It was so obvious that nobody had seen it.

The cock does not feed his mate on the nest, nor have I seen him incubating eggs.. But once in my garden he was brooding small young.

THE COAL TIT.

(Parus ater britannicus).

Field characteristics.

Its small size, sooty appearance, and white nape distinguish this tit from others of its genus.

Allied species.

In the Irish form ,the white is replaced by pale yellow.

Song.

Loud and clear and quite distinctive from the other tits. In two syllables:—Too-hee, too-hee, too-hee.

Call and alarm notes.

Described in the ' Handbook of British Birds ' as ' Tsee-tsee-tsee.

Eggs.

Seven to twelve. This is often an early breeder among the tits in the South, the 15th April will find many hens sitting, except possibly those in nest-boxes. These owing to competition may be delayed 10 days or so.

In my opinion most Coal Tits are double-brooded in the South. Another clutch of 5 or 6 eggs is forthcoming about mid-June.

D. W. Musselwhite considers that 15 days is the normal time needed by this bird to replace a lost nest and eggs.

Nest finding.

The Coal-Tit is very fond of holes in banks. Sometimes I have noticed the nest-hole when walking along, by the smallpiece of rabbit-fleck sticking to the side of the hole.

One must listen for the clear call of the cock, and follow him up. He feeds the sitting hen very frequently and can easily be watched to the nest. Also she will come off and join him, when he will also feed her. She must be watched back, as she seldom stays away from her nest for long. The hen alone incubates. These birds feed usually very near the nest site, and only leave the vicinity to drink. Therefore I consider it rather an easy nest to find. Whether it can be inspected when found is another matter. In the pine and heather country where I live, the nest hole is mostly in sand, and any interference with it causes a fall from the roof of the tunnel with a consequent smothering of the nest.

In the West of Ireland I once found a nest of the Irish Coal-Tit with eight eggs, in the broken off and rotten root-stub of a small pine tree. I had watched the hen return to her nest, which was about six inches below ground and directly beneath the entrance, so that one looked down straight on to the sitting bird. What happened when it rained, as it does most of the time in the West of Ireland—I cannot imagine.

Carlyon-Britton once saw a Coal-Tit taking moss into its nest-hole when it was brooding on well-incubated eggs.

THE COMMON SANDPIPER.

(Actitis hypoleucos).

Field characteristics.

A small wader—length about 8 inches, brown on upper parts and white underneath.

Allied species.

None with which it could be confused in the breeding season.

Song.

A variation of its call notes.

Call and alarm notes.

A loud clear ' Tu-hee-hoo,' repeated several times.

Eggs.

Three to four. I have once found five. Single brooded. Will replace a lost nest and eggs in 13 days. Both sexes share in incubation. The week commencing 18th May is the best time to find eggs.

Nest finding.

This Sandpiper loves to nest near water, and the favourite site is undoubtedly a river bank.

It does not normally breed in the South, and is confined to the North and West. Some writers consider it a close sitter. My own experience is that about half the pairs met with, sit closely and can easily be flushed by walking the river banks. The other half do not. These birds leave eggs when one is about 100 yards off, they then make straight for the water, and fly off silently and low, hugging the near bank, and so in a matter of a second or two are out of sight. It pays to work up wind. The non-sitting bird may be 200 yards or more from its mate. This one will usually rise silently, and move but a

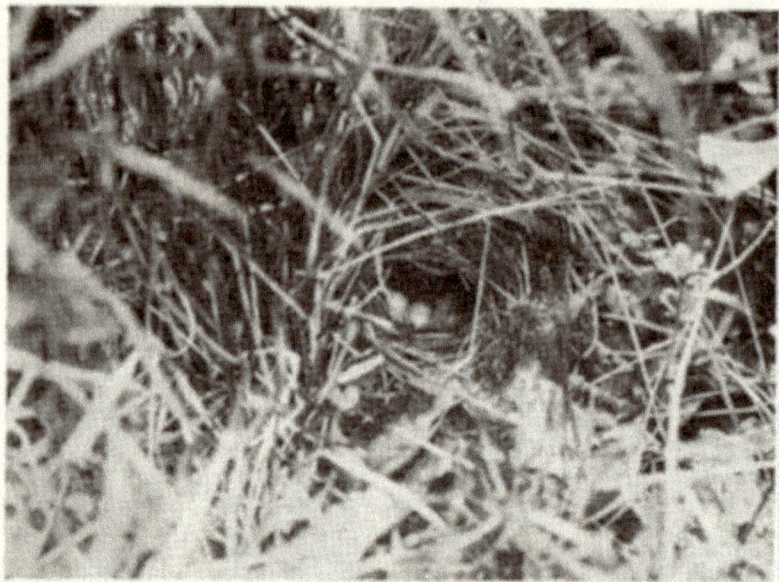

NEST OF WILLOW-WARBLER.

Photo by Rev. C. J. Pring.

NEST OF WOOD-LARK.

Photo by Rev. C. J. Pring.

NEST OF CIRL BUNTING AMONGST FOLIAGE GROWING INTO BRIAR BUSH.

Photo by G. Tomkinson.

NEST OF CORN BUNTING.

Photo by N. B. Coltart.

short way before alighting. On the other hand, the sitting bird if put up when off eggs, calls and appears agitated. Now is the time to watch it back. Owing to the winding of the river banks they are difficult to keep in sight. If one is too close they will not go back, and if too far away they cannot be kept in sight. One consolation is that they do not stay off eggs for more than about five minutes before attempting to return. I have timed them back many times. They do not fly straight to the nest, but either alight at the water's edge opposite, and pretend to feed, or perch on a branch for 2 or 3 minutes before flying down to within a few feet from the nest, when they run on to it.

One bird I watched back on the River Forth in Scotland, sat for several minutes on a dead bough sticking out of the water. I had her in my glasses at 150 yards range. Then she flew inland and settled on a wire fence. Here she waited for another two minutes before dropping down on to a grassy bank. I had the spot marked exactly by a dead stub hard by. I gave her 5 minutes to settle down and walked to the spot. Not a sign of a nest was to be seen, and I eventually flushed her off eggs, twenty feet from where she dropped.

THE CORN BUNTING.
(Emberiza calandra).

Field characteristics.

A plump lark-like bird, with a rather long tail, both sexes are alike and have dark bibs, but the male is noticeably the bigger. Like the Red-backed Shrike this bird when perched seems to carry its tail lower than other birds. In flight the legs are often allowed to droop.   c

Allied species.

None.

Song.

Described as ' jangliug keys ' or ' pounding glass in a mortar.' Neither of these in my opinion gives a true description. The nearest I can get is: ' Tic-tic-tic-tic-tic-tic-tic--chee-ee-se.' The sound, however, carries a long way, and it is easy to pick up the singer, as he favours the top of bushes, fences, wires, hay-stacks, roofs, and the tips of growing crops.

Call and alarm notes.

The alarm is a loud ' Chip-chip-chip ' with a good pause between each chip.

Eggs.

Three to six.

Double brooded in the South, if not treble-brooded sometimes. The last week in May in the South for first broods, and mid-June in the North, is about right for fresh eggs. But this bird is very un-certain and nesting depends on the state of the crops. The second brood is ready from about the first week of July. I know of fresh eggs found in August.

It will replace a lost nest and eggs in 10 days.

Nest finding.

Whether this bird's nest can be found depends on what sort of vegetation is chosen by the hen for her nest. If in growing corn, the position is hopeless, as even if the hen is seen to return to the nest, one cannot go in to investigate without incurring the wrath of the farmer.

But if the site is open downland, grass fields, or gorse bushes then I do not think that it is very difficult to find, if you have the necessary patience. The cock is somewhat addicted to polygamy, and this

helps, as he has probably got more than one hen
sitting in his territory. The hen Corn Bunting leaves
her nest to feed once every hour, with great regularity,
and the cock goes with her, and brings her back after
feeding, if not right to the nest, at least a great part
of the way.

The last two hours of day light will see great
activity over the breeding ground.   Then the hen
leaves her charge more frequently, and this is the
best time to find nests. Watch the cock, he will see
her leave first, and hurry after her. They feed right
out of sight from the nest. The next thing to be seen
is, both returning together at great speed.   Sometimes they will hurtle round and round the area, just
skimming the ground, the cock in pusuit of his mate,
who turns and twists like a driven Snipe. It is with
difficulty that they are kept in the vision of the field-
glasses.   Suddenly the cock breaks away and alights
on one of his perches and bursts into song. This is the
moment—the breakaway—when sight is so often lost
and the cock followed instead of the hen.   If that
happens the hen has plunged on to her nest, unseen,
and will have to be watched off once more.   Sundry
House-Sparrows and Skylarks flitting about, cross the
trail and help to make confusion more confounded.

If, on the other hand, the hen has been kept in
sight, then her destination will have been noted, and
it will only be necessary to walk up and tap around
with a walking-stick, when up she rises with a loud
whirr.

Some hens return to nest rather differently, by
first perching on some prominent mark such as a
thistle or dock, near the nest. She then looks all
round, sounding the alarm and when satisfied that the
coast is clear, drops on to the nest.

I have found the period from about 1 p.m. to
4 p.m. B.S.T. is a very bad time for watching these
birds.   Little, if any, movement or feeding takes

place then, and I think one can employ one's time more profitably by a little judicious beating. The chief singing posts of the cock will have been noted, and one of them may be close to the nest. Tap out likely looking spots and try to flush the hen.

Once on the South Downs near Worthing I noticed a Corn-Bunting singing on a clump of gorse on a slope of a hill. Below him was a field of coarse grass with a few brambles and small hawthorn trees. The nest will be in the grass for sure, I thought, but as I passed the gorse I gave it a shake with my stick where it was thickest. An amazing sound of squawks, like a Starling when handled, came from deep down in the gorse—but nothing came out. Mystified I parted the branches with my stick, and there, deep down inside, was a Corn-Bunting's nest with four eggs, the hen standing on its rim, apparently unable to get out. She flew off at once, none the worse, but I have never heard a Corn-Bunting make that noise since.

THE CROSSBILL.

(Loxia c. curvirostra).

Field characteristics.

About the size of a Greenfinch with a very thick bill, the male scarlet, the female greenish-brown. In the field the hen except for her green rump, looks like a stout Sparrow.

Allied species.

Not distinguishable from Scottish Crossbill in the field. The Two-barred Crossbill has white wing-bars.

Song.

Described as like a Starling's by most writers, but I can hear little resemblance. On the other hand, the Starling can imitate the Crossbill's song exactly.

The sound can be described as a series of rapid ' Schip-schip-schip-s,' in volume and sound not unlike the alarm call of the Blackbird, but more pleasing to the ear, and interspersed now and then with some quite melodious notes.

Call and alarm notes.

Call—a loud metallic ' Chip-chip-chip,' used by both sexes and invariably uttered on the wing. Alarm—A rapid edition of the call note, sounded very much more quickly and not so loudly.

Eggs.

Two to five. Usually 3 or 4. Most birds are single brooded. Almost impossible to suggest with accuracy any one date when eggs could definitely be found.

I find that I must draw a distinction between our resident birds, and those that come to us in large numbers from the Continent in irruption years.

Resident Crossbills.

In the South they appear on their breeding grounds towards the end of February, and the middle of March, 13th-17th is the most likely time for incubation to commence. An odd nest might be found at the end of February, and I have known a pair only starting building early in April. But March is the best month for eggs. Nests have been found with eggs in December in Norfolk, and I once found a nest here on 9th February, from which the young had just flown. This means that incubation must have started early in January.

Immigrant Crossbills.

These behave in a manner different from our resident birds. In the first place only a percentage actually nest. Flocks of adult Crossbills, the males in scarlet plumage, and singing vigorously, frequent

the district where their friends are nesting, yet themselves make no attempt to do so. This mystery has never yet been explained.

I am inclined to the theory, and it is only a theory, that our immigrant Crossbills nest later than the residents.

I divide their breeding into three main waves or phases:

        1st. Mid-March.
        2nd. End-March.
        3rd. Mid-April.

One nest that I found on 5th May, I believe to be a genuine second brood, but I am unable to prove it.

The normal time for the replacement of nest and eggs is 10 days, but I have known it to take 12 days. If the new nest is close to the old one, in the next tree as is often the case, the latter is pulled to pieces and used in the construction of the new nest.

For practical purposes the Crossbill is single brooded, at least this is my opinion. If they are double brooded it is strange that after the young are on the wing, there is no sign or sound of Crossbills in their breeding haunts, except intermittently and then only in flocks.

The Crossbill is a rapid nest builder, in fact I know of no small bird that spends less time over its first nest. Therefore if one wants to see them at work it is necessary to be about during the best five days in the year 1st to 5th March and put in a lot of hard work .

If Crossbills are about in the district their traces may be found in ' worked ' cones, see diagram opposite, near the base of Scots pine and larch.

Ears should be kept open for the cock's song uttered from the tip of a pine, as he often sings near the nest, or even on the same tree. He

accompanies the hen on her building forays, the
material is usually gathered from the ground
near the nesting tree, in which case the cock
either follows the hen almost to the ground, or
watches from the top of a convenient tree, when
he will sing his loudest.

When the time comes for the final feathers to
be added to the lining of coarse grass, the procedure
is rather different. It may be a longish way to the
chicken-run the hen has set her heart on.   With
loud ' chip-chips ' they fly in a bee-line, right out
of sight.

LARCH CONE

SCOTS PINE
CONE SHOWING
LACERATION

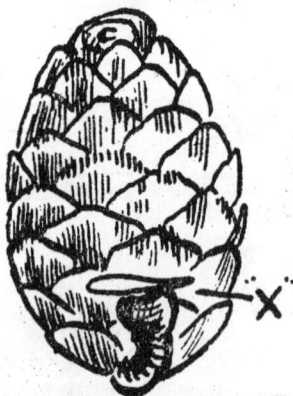

SHOWING 'CUT'
CLEARLY
AT "X"

Stance should be taken up where they were last sighted and their return awaited. It may be 20 minutes or more before they re-appear, a large feather in the beak of the hen. (Occasionally the cock will carry feathers too). In this way with patience the source of their supply may be traced, and the cock will be seen mounting guard, while his mate is hopping on the ground choosing feathers. I have noticed that Sparrows seem strongly to object to the removal of 'their' feathers, and the hen Crossbill has to run the gauntlet of angry pecks. If then, Crossbills are about, but the nesting site is elusive, a visit to the neighbouring chicken-runs might be profitable. The observer can watch these birds extremely close by, as they show no fear of man.

It is possible to walk past a nest of Crossbills day after day and never see a sign of the owners. In an area where there are thousands of pines and birds few and far between, I consider this nest a difficult one to locate, and to do so needs real hard work and a certain amount of luck. Adding to the difficulty is the dreadful weather often encountered at this time of year. A high wind makes it almost impossible to hear anything. Now and then a nest may be discovered by thoroughly inspecting through glasses every branch of each pine, round the singing places of cock. I have found them by seeing the cock feed his sitting mate, this is done in a flash, and may easily be missed, especially as some nests are placed high up and are invisible from the ground.

The curious habit of these birds of fouling the nest itself, and the branches around, even before the eggs are laid, has been also noted by N. Gilroy in the 'British Birds' magazine. I have found a nest by seeing this 'whitewash.' There is one note which is uttered by the cock Crossbill when breeding or about to breed I can only describe it as a

squeaking rattle. It is quite distinctive from its normal song, and if heard, invariably means that there is a nest to be found. I have never known it to fail.

During breeding and mating, some of the displays by the cock are very beautiful. He will parachute between two trees like a cock Tree-Pipit, giving the ' breeding rattle,' and if you are lucky enough to see him then, with the sun-light on his glorious scarlet plumage, the memory of the sight will never fade.

Crossbills are sociable birds, and even when paired and nesting, love to flock together for feeding. I have seen a cock leave a large flock flying over, and drop into a tree, where I soon discovered he had a sitting mate. A strange thing about these birds is their love for certain favoured spots to nest. There are thousands of pines around my home, all much alike to human eyes, yet year after year nests are found in the same old spots. I have spent hours and even weeks patrolling miles of similar country with only feeding parties to be seen. It is said that they like to nest by a road, they often do, but two of my colonies are far from one.

I have not seen the cock incubate; this starts from the second egg laid, and the hen is then seldom off the nest. On several occasions I have watched a pair choosing the site of their nest. In and out of the tree they go, creeping about the branches like tiny parrots, yet always returning to the same spot, calling loudly the while. At last the hen breaks off with her powerful beak a small twig from the nest tree, and when once this is fixed in position, no more time is wasted and building proceeds rapidly.

On the 26th March, 1936, the year of the great Crossbill irruption, at 1.45 p.m., I witnessed a strange sight. I saw, what I thought were two

Sparrows sitting side by side on a telegraph wire by the roadside. As I came nearer two cock Crossbills in brilliant red plumage, flew and settled one on either side of them. Through glasses I now saw that all four were Crossbills, the inner two hens. Both cocks turned inwards and fed the hen next him, who fluttered her wings and jerked her tail like a young bird does when fed by its parent. Both hens then dropped to the verge of the road below, and tearing off a beakful of grass, flew straight to half-completed nests in adjacent pines not 50 yards away.

THE CUCKOO.

(Cuculus c. canorus).

Field characteristics.

Blue-grey above, with white underparts barred with black, about the size of a Turtle Dove. The hen is browner on the back, and this feature is discernible in the field. In flight the bird is unmistakeable. Its long tail, rapid direct flight with quickly moving wings make it hard to confuse with any other bird, even at a great distance.

Allied species.

None.

Song.

The well known ' Cuck-oo ' needs no description.

Call and alarm notes.

Both sexes have a curious call described by E. P. Chance in the 'Cuckoo's Secret' as ' Wha-wha-ing.' This call I think is a sign of sexual excitement, and reminds me of a throaty cough. I have never heard it after laying is finished. The hen has in addition a liquid bubbling note, also mentioned by Chance. This is uttered immediately

after egg-deposition. I have often heard it when
the hen is flying over her breeding territory,
watching her victims, and on leaving their nests
after one of her many visits of inspection.

It is something like a Dabchick's cry, but less
shrill.

Eggs.

E. P. Chance has proved that the maximum
number of eggs a Cuckoo can lay is 25 in a season.
But this record number could never occur in the
natural state. In this case the fosterers (Meadow-
Pipits) were regularly ' farmed '* in order that a nest
should be ready for the Cuckoo on each of her
laying days throughout the season. It is to be
doubted whether Cuckoos lay, more than 8 to 10
eggs under normal conditions, and many must lay
less, especially birds in their first season.

A Cuckoo can lay only every other day, and
then only if it knows of a nest in a suitable condition
to receive its egg. That is to say the eggs in the
nest must be fresh and not incubated.

By far the majority of eggs are laid in the first
half of June, but eggs can be found from about the
second week of May to the 24th June, or even a
day or two later.

* The eggs of some of the fosterer Pipits were removed.
This would compel them to build and lay again, so that there
would always be a new nest with one egg on the fifth day
after being taken.

Nest finding.

Boys have a keen ambition to find a Cuckoo's
egg, but the egg itself is of little value unless there
is with it an egg or eggs of the fosterer in whose
nest it was discovered. The reason for this is as
follows:

It is a generally accepted hypothesis that the hen Cuckoo lays her egg in the nest of the bird in which she herself was reared.

Therefore, there are races of what may be described as Meadow-Pipit-Cuckoos, Hedge-Sparrow-Cuckoos, Robin-Cuckoos and so on. So naturally it follows that a Cuckoo's egg by itself is of no scientfic interest. Now we know that the Cuckoo does not reach its breeding ground until late in April, when most of our resident birds are still busy with their first broods, and the migrants have not yet started nesting operations. So, the number of nests of her own particular victim which she may be able to find is limited, and that is probably the reason why so few Cuckoos' eggs are found early in May. I remember one very late spring, the Pied-Wagtails in Sussex were so late nesting, that they received the Cuckoo's egg in their first nest in mid-May, instead of in their second in June as is usual. This meant that those victimised, raised no young of their own that year.

Taking the British Isles as a whole, two birds stand out far above the rest as favourite hosts for the Cuckoo's egg.

Meadow-Pipit   (Resident)
Hedge-Sparrow   (Resident)

next come

Reed-Warbler   (Migrant)
Pied-Wagtail   (Resident)
Robin   (Resident)

followed by

Sedge-Warbler   (Migrant)
Tree-Pipit   (Migrant)

In the nests of these seven birds, will be found the vast majority of Cuckoo's eggs laid in these Islands.

A few eggs may be found in nest of

Linnet (Resident)
Yellow-Hammer (Resident)
Spotted Flycatcher. (Migrant)

Of the five chief hosts then, the Hedge-Sparrow, Reed-Warbler, Pied-Wagtail and Robin can be found only by assiduous searching, with perhaps the exception of the Pied-Wagtail—as this bird can be watched when it favours farm-buildings, haystacks, etc., for its nesting site. There will be many disappointments. Even if there is near home, a Cuckoo parasitic on Hedge-Sparrows or Robins, it will be found that many nests of these common birds do not contain a Cuckoo's egg. Hope must not be lost, and the search kept up each season. It will be worth all the hard work when success is achieved.

Now for the Meadow-Pipit. This is the easiest fosterer of them all to work, it likes open spaces to nest in, and here it can be observed. I start about 25th May, by strolling over the ground, and find a few nests by watching the birds. Then I see in what state the majority are, i.e., feeding young, in or out of the nest, building, etc. This will depend on what kind of a season it is, early, average or late, and I make my plans accordingly.

If Cuckoos are working the area some sign of them will be seen or heard. They are partial to perching on telegraph wires or the posts. From this vantage point they command a fine view of the ground. This is good from the watcher's point of view as they can be picked up through the field-glasses, whereas when in a thick tree, are invisible. If the bird is disturbed when on one of her perches and flies off, the odds are she will be back again in a few minutes. So a position should be taken up, not too near, and her return awaited. She may then sit motionless for long periods—watching, or she may drop to the ground and swallow a caterpillar

and return to her post. Sooner or later—usually
later!—she will visit the nest she has been watching.
If in long heather, etc., she may fail to find it, even
after a lengthy stay on the ground. She will be
continually buffeted by the angry Pipits, till she
loses all sense of direction, and several breast
feathers too. I have seen one of the Pipits sitting
on her back, in an effort to drive her away. After
failing to reach the nest she will return to her tree
many times until successful. It is not always wise
to go over and look for the nest directly (though
the temptation is great), unless you want to see
her lay her egg. Then it is essential to know when
the Pipit starts laying. Otherwise it is safer to
wait five days, when there is less chance of
desertion, and the Pipit may be flushed, which is
a great help.

Some nests of Meadow-Pipits are extraordinarily
well concealed, and will not be found without
flushing the bird. A search may end in its being
trodden under foot. I remember one day in early June,
I was watching a hen Cuckoo that I knew well. I
saw her lay her egg in a Pipit's nest in long heather.
The position of this nest I knew beforehand, as the
Cuckoo herself had shown it to me a few days
before. Before she laid, I saw her visit another pair
of Pipits, the whereabouts of whose nest I did not
know. I walked to the spot and flushed a Pipit off
a nest containing one egg of the owner. That
evening I telephoned a friend and asked him if he
would like to see a Cuckoo lay its egg in two days
time. He replied that he would be delighted and
could he bring a friend? To this I agreed. We were
to be on the ground at 3 p.m. B.S.T. The day
arrived, and we sat in a car about fifteen paces from
the nest. There was the Cuckoo watching from a
telegraph wire 150 yards away. After a few minutes
had passed she made her long glide and landed in

the heather near our car, but not very near the nest.
The Pipits attacked at once, and she could not find
the nest. She must have made more than a dozen
glides from the wire, and each time something upset
her. An old cock Blackbird set about her once and
drove her off. It was not until 6 p.m. that she
found the nest, removed one egg in her beak, sat
down—laid and flew off bubbling. We walked across
and there was her egg still warm, and two of the
Pipit's eggs. My friends thought that this was
marvellous, but it was really very simple:—I had
seen her lay once, and I knew she could not do it
again until 48 hours had passed. I knew that there
was a nest with one egg ready for her, which she
had shown me. I also knew that this particular
bird did not lay before 3 p.m. Therefore it was
long odds that she would use this nest in two days
time;—she did.

E. P. Chance believes that the hen Cuckoo
will take possession of an area, and will drive off
all other hens bent on laying there. He calls this
the ' dominant ' bird. He also states that the
dominant bird will prevent the others from laying
until she herself has finished for the season.

My own observations do not confirm this
theory. No less than four hen Cuckoos were working
one Common which I had under observation for
long periods during several seasons, though I admit
that there was occasional fighting amongst them.
They often drove one another from their observation
posts, but they only moved to another some way
off, and returned when the coast was clear.

Once I saw a hen go down to lay, and she was
attacked by another whilst on the ground near the
nest. They had a grand battle, and the heather was
covered with feathers from their breasts. Neither
bird managed to lay that day at that particular spot,

NEST OF YELLOW WAGTAIL.

Photo by Rev. C. J. Pring.

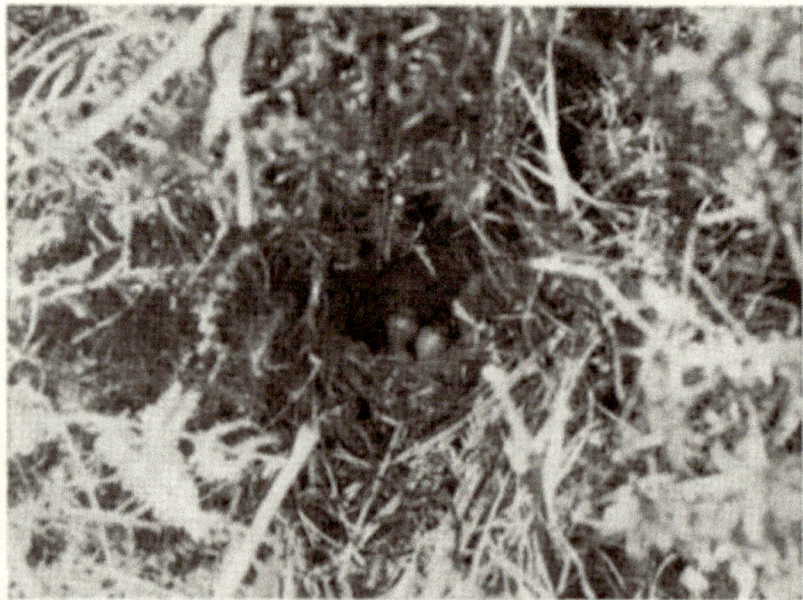

NEST OF STONECHAT.

Photo by Rev. C. J. Pring.

NEST OF COMMON SANDPIPER.

Photo by Rev. C. J. Pring.

NEST OF SNIPE.

Photo by Rev. C. J. Pring.

but the fact remains that I found eggs from all four Cuckoos on the Common, sometimes two in one nest, and on two memorable occasions three in one nest. So it does not look as if the ' dominant ' one did her job very thoroughly.

A short note appeared in the magazine ' British Birds ' under my name, in which was described how on another Common there were two hen Cuckoos working together in perfect harmony. I found eggs from both these birds together in the same Pipit's nest. These two birds used the same lookout post, sitting side by side. When in flight, one followed the other like a dog, and settled beside her whenever she alighted.

All these points go to prove how risky it is to generalise in nature.

The Cuckoo has always fascinated me, and I devote much time each season to its study. The foregoing is not intended to be a treatise on this bird, but merely a few hints on what may be expected, and how to set about the study of the Cuckoo.

THE DARTFORD WARBLER.
(Sylvia undata dartfordiensis).

Field characteristics.

A small bird about the size of a Wren, blackish-brown above, with reddish-pink breast, and a very long tail.

The hen is duller and browner.

The cock at close quarters with the sun on him is very beautiful, his plumage then seems to reflect the vivid tints of flowering heather, amongst which he lives and has his being.

Allied species.

The Whitethroat is the only other bird with which the Dartford can be confused. But as its

d

name implies its throat is pure white and underparts whitish.

Song.

A sweet clear warble, reminding one of the Hedge-Sparrow and Stonechat's songs, but purer. It sings on top of a bush, etc., so is quickly identifiable. A persistent singer in every month of the year on mild days with no wind.

Call and alarm notes.

The call is a harsh ' churr,' also used as an alarm note by both sexes. The hen's is never so loud or harsh as the cock's and with practice is soon distinguishable from his. It is not unlike the White-throat's scolding cry, but softer and less grating to the ear. The alarm may be sounded at any time of the year, but as the hen is much more shy than her mate she is not likely to be heard so often, unless the nest is approached with young in it. When the cock is really aroused he bobs up and down with excitement and his notes then sound like ' tues-dee, tues-dee-dee.'

Eggs.

Two to five. Normally three or four.

The earliest Dartfords in a mild Spring start to sit on 16th April. I have heard of eggs in March in this district, but have no personal experience of them. Nest building by early pairs may start the last half of March, but these birds have a curious habit of with-holding the laying of eggs after the nest is completed. This period may be anything from five days to as long as a fortnight, and is to be expected during a cold spell which so often occurs in an English Spring.

The last half of April will find more birds laying, although during the last few cold Springs most birds did not start to sit until the second week in May.

The year 1936 was especially backward, and the earliest bird that I knew then, did not start laying until 29th April. The majority began incubation about 17th May.

The Dartford Warbler will replace a lost nest and eggs in from 10-13 days. Double-brooded.

Nest finding.

A good deal of ink has been expended on this subject of recent years. Walpole-Bond in ' Field studies of some rarer British Birds ' devotes a chapter to the Dartford Warbler as observed on the Sussex Downs, and to the difficulties he has encountered in searching for their nests. That writer has told me that he considers it harder to find in gorse, such as is found in quantity on the South Downs, than in heather which is its chief habitat here in Hampshire.

Now what is the best way to set about finding a Dartford's nest?

(a) By watching the birds building.

Both take part, the cock often does a little on his own, but he usually accompanies the hen when she takes material to the nest. If the watcher is unseen by the birds no difficulty should be met with, but it is worth remembering that they seldom show themselves more than they can help, and that though they fly direct to the neighbourhood of the nest, they complete the last lap by creeping through thick cover if they can. Also they have a disconcerting habit of over-shooting the nest, and creeping back to it unseen, so if this trick is not known much time can be lost in searching in the wrong place.

It is much better to mark the place where the bird comes away from, after it has disposed of its load, than to search around the spot to which it appeared to go. Dartfords are uncertain builders,

and spend long periods away from the nest site, and although building may be noted at any hour during daylight, one is just as likely to see no sign of activity.

(b) Flushing a sitting bird.

Other writers have stated that this bird can easily be flushed once incubation has started. It can of course, if you know where the nest is! But when a survey is made over a huge expanse of heather and gorse, all equally possible, where is one to start?

Full of optimism the seeker plunges into the jungle, tapping with a stick likely clumps, and even searching in the gorse here and there. If successful thus in flushing a sitting bird, luck is indeed in. One is more likely to find nothing but a Linnet's nest or two. Think of the odds against achievement, for even if working on a known breeding site, one has to contend with the fact that at least a month may separate the early from the late nester, so that they may not be incubating at all.

Now and then at long intervals a nest does turn up this way, but I certainly consider indiscriminate beating practically useless.

First one must find the birds, and I suggest a walk over the area in strips about twenty yards wide. These birds have a habit of resting occasionally between building periods, tucked away out of sight right beside the nest. If disturbed they fly off together and plunge into a thicket, when the place from whence they came should be marked with a hat, stick or handkerchief. A brief search will soon settle the point. I personally have found a number of nests in this way, before an egg was laid.

(c) By rightly interpreting such clues as the cock may give.

The cock Dartford Warbler has several favourite singing or display posts and on the approach of a trespasser into his territory is likely suddenly to appear on the top of one of them, and scold or even sing. These posts are not very near one another, but one of them is certain to be close to the nest. The question is how to find which one, for it is of little use searching unless it is done slowly and thoroughly, which takes time. I suggest that if there is a patch of gorse it should be parted with a stick to see if it is used as a roosting place, this is quickly noted as a large accumulation of droppings is not hard to see. If no sign, look out should be kept to see whether he repeats his alarm some distance off; if so another try may reveal the roost (of course it may be in long heather). The roost may be near the nest.

It happens that after scolding for some time, he will fly right away out of his territory, and unless he is seen to go—which is not by any means certain if this ruse is not known—the searcher finds that all is quiet and not a bird is to be seen. This may mean that the hen is safely on the nest and that one is not near enough to cause him any anxiety. It might pay to move and try places a bit further on in the hope of flushing the hen. Personally, if the cock vanishes, I prefer to leave his territory and return in twenty minutes, when he should be there to greet me.

I have tried to describe what happens in Dartford territory when the hen is sitting, at which time the searcher has the best chance of success. If both birds are off feeding the cock will bring the hen back quite openly right to the nest. When the cock takes over the duties of incubation he may fly straight there from fifty yards, and I have walked direct to an unknown nest in this manner, after having spent hours watching and searching in the

wrong places. But once he is safely on, then the trouble begins, for this in my opinion is the most difficult period, as no clues are forthcoming.

The hen is silent and invisible, and one would say that there was not a Dartford within hundreds of yards. I once found a nest in a curious way. Walking home after a tiring and blank day, I heard the alarm note of a cock Dartford, uttered once, from the hill-top above me. Something seemed to urge me on, so I ran up the prickly slope and sat down on my shooting-stick in a small bare patch over the brow. Not a bird to be seen, and long heather everywhere, a hopeless outlook. Suddenly out of the heather ten yards ahead appeared for a second, a hen Dartford, scolded once softly and disappeared.

Rising quietly I pulled up my shooting-stick and tapped the nearest tuft of tall heather. A cock Dartford fluttered off a nest of four eggs!

I had read that slender clue aright.

What I think happened was this: The cock was returning to the nest, to take over incubation from his mate. He caught sight of me as I walked through his territory, so scolded once, the change-over taking place when I was walking up the short slope. The nest was just over the brow, and the hen after leaving her nest, saw me and came back to warn the sitting cock.

• The reader must not imagine that by following these hints he will find every nest of this species that he looks for—far from it. I consider it really is a most elusive one, and a lot of time will have to be given if success is to be attained. And even then only a small proportion of the looked-for nests will be found.

THE LESSER REDPOLL.
(Carduelis flammea cabaret).

Field characteristics.

This little bird should be easily recognised in the field by its small size and deeply forked tail. In general appearance it resembles a Linnet, but is considerably smaller and has a red forehead and black chin.

Allied species.

Its near relative the Mealy Redpoll only occurs with us as a winter visitor and is larger.

Song.

The song of the male is very distinctive, and I have yet to read a good rendering of it. It consists of four sounds — ' chee-chee-chee-chee ' quickly uttered and followed by a loud rattle. During the breeding season he is very noisy, wheeling high in large circles, singing over his nesting ground. He will dive to the top of a tree and continue his song from there. He may be joined by another cock and together they race through the sky—round and round, rattling away like little alarm clocks.

Call and alarm notes.

Call note is like a subdued version of the ' chee-chee ' in the song, but slower and softer. Alarm is a plaintive ' whee-ee,' to my ear indistinguishable from that of the Goldfinch.

Eggs.

Three to five. I cannot give a certain date for finding eggs of the Redpoll, I have been misled too often myself. They are fond of nesting in small colonies and all do not lay at the same time even within the colony. Generally speaking, it is a late nester, the first half of June should be the best time, but some will be hatched before the end of

May. A few are double-brooded, laying again early in July. The nest is worthy of mention as the text books say that a foundation of twigs is characteristic of the species. I have found nests without a single twig in their composition. Some are bulky and untidy, others the reverse, the latter can be so small that when tucked away in a fork of birch or juniper are invisible from below.

Scottish eggs are paler than those from Lancashire.

Nest finding.

Redpolls' nesting habits vary according to which part of the British Isles they occur. My experience of them runs to three widely separated areas which I shall endeavour to describe from North to South.

Perthshire.

The most favoured site here is undoubtedly in Juniper. This bush occurs in scattered clumps on the foothills and mosses of the Western Highlands. Here and there one finds a juniper that has ceased to be a bush and become a good sized tree. Nests may be found in either, high up on the branches or as low as 3 feet 6 inches in a straggly bush, like a Linnet's nest in gorse. I found one nest far from the rest of the colony, this was on a small rocky hill-top. The juniper was thin and the nest was on a bare fork. The sitting bird was obvious from a distance of quite fifteen yards. But as a rule in juniper they take a bit of finding. The cock is certainly a help here, and it saves much fruitless searching if one sits down overlooking the area for a quiet watch.

Redpolls have a habit of flying right away from their nesting area at irregular intervals, and it is of little use to follow them up. When the cock returns,

often in company with another cock, they will separate when over their homes, and dive each to his own area. This is the time to mark down the singing post of each. If the hen is with him it is much easier as she always sounds the alarm note when approaching her nest. The hen sits very closely and is almost as tame at the nest as a Crossbill. They are audible in flight a long way off, so it should not be difficult to pick them up. But there are complications to be met with in the form of young out of the nest. These are noisy and restless, but can be picked out through glasses by the absence of the black bib under the chin. After juniper the next vegetation in favour is birch. A typical site is against the trunk amongst the suckers that grow from it. Here most nests are easily seen as they are seldom higher than 18 feet, but an odd nest may occur high up on a lateral branch of a large birch and be quite inaccessible.

The last site I propose to describe is in village gardens. Here the apple-tree is the chief attraction. I once watched late in June two nests being built in adjacent trees, the branches of which nearly touched each other. The young of the previous broods were in close attendance and kept pestering their mothers for food much to their annoyance. The noise was amazing, so much so that the little garden seemed to be swarming with Redpolls.

Lancashire.

The silver-poplar is easily first favourite in this part of the country. It is found in little clusters, too small to be called woods, on the out-skirts of towns, golf-courses, etc. This tree is one of the last to come into leaf, and is quite bare until the middle of May or even ten days later. The early Redpolls do not wait for the leaves to come out, but build their nests in the suckers against the trunk, or even

on the bare branches. It is surprising how one can walk past an exposed nest, such as I have just tried to depict, without seeing it. But I have done so frequently. I mentioned once to a friend who lived on the spot, that a certain group of silver-poplars near his house seemed to be popular with Redpolls. He told me that two pairs of them nested there every year. I went through the place very carefully and found nine occupied nests!

After silver-poplars come wych-elms. There are plenty of these in the towns and suburbs. The nest is usually high up, and hard to see until the leaves come off. Some gardens have a hedge of wych-elms which are kept pollarded. Redpolls love these and the tiny nest is easily overlooked.

Alders round meres is another place in which they should be sought, and there is an odd nest now and then in a hawthorn.

Hampshire.

The behaviour of Redpolls in the district round my home is somewhat different from those in the North. Here, although huge flocks may be seen in March numbering 200 or more, most of these have disappeared by mid-April. This, in spite of the fact that some members of these flocks may be seen mating, feeding hens and otherwise showing signs of breeding activity.

Our resident Redpolls are very thinly distributed, chiefly in the pine and heather country. They breed in scattered pairs in gardens where there are large spruce, firs, larch or pines. The nest is normally high up on a lateral branch.

I do not consider that it is double-brooded in the South. If it loses its eggs and nest they are replaced in ten days, and the old nest may be pulled to pieces and used in the construction of the new.

THE MARSH TIT.

(Parus palustris dresseri).

Field characteristics.

A small black-capped tit whose cap is shiny black.

Allied species.

See under Willow-Tit.

Song.

A monotonous repetition of one note, loud and penetrating—' chee-chee-chee.' It does not sound much like a song at all but a noise. He has in addition a spring song which incorporates the notes of the Redstart and Whitethroat and is quite musical.

Call and alarm notes.

The call is a noisy ' titsoo-titsoo,' which is incorporated into the alarm note, this is a harsh ' chic-a-bee-bee-bee-bee.' This description is by Walpole-Bond.

Eggs.

Five to ten. Most Marsh Tits are sitting by 6th May in the South in an average season. My earliest record is a c/9 on 22nd April. Single brooded. Clutch replaced in about 12 days.

Nest finding.

The hints that have appeared under Coal-Tit could equally apply to the Marsh Tit. But the Marsh Tit's favourite site for its nest is a hole in a tree. Holes in banks and walls are used too, but trees are preferred. On occasions this tit will bore its hole itself after the manner of the Willow-Tit. A North Easterly aspect is preferred, but I have found nests in banks facing South.

This bird is about the closest sitter of the

whole tribe of tits, and sometimes nothing will make the hen leave her nest. If one wishes to inspect its contents this can be very annoying. I have found a method which may sometimes help: This is to sit down near the nest and wait for the cock to appear on the scene. If really close he will at once sound the alarm. One should walk over to the nest-hole and look inside afterwards retiring a little way off, but not too far, then he may call the hen off her nest. The only times I have known the Marsh Tit to flush easily is when they have been situated in rotten timber.

THE NIGHTINGALE.

(Luscinia m. megahyncha).

Field characteristics.

Chestnut-brown in colour, pointed wings and a long tail distinguish it in flight, when it is reminiscent of a small reddish thrush.

Allied species.

None.

Song.

Most country dwellers know this song even if they cannot recognise any other. It usually starts: ' jug-jug-jug-jug-jug ' rapidly repeated. A pause—then—'chuck-ee-chuckee-chuckee jug-jug,' etc. The famous crescendo may be worked in anywhere, this is ' pee-pee-pee-pee-pee-pee ' getting louder and louder, then a pause; the above may be repeated in any order.

The song is seldom continuous by day, and is broken up by many long pauses.

Some cocks never seem really to get going and may be rated as poor songsters.

Nightingales are at their best at night, when there are several cocks singing against each other, then it can be understood why no other bird can even approach their volume and beauty of tone.

Call and alarm notes.

The alarm is a strange growling note, very distinctive, and used by both sexes. I have known a cock to stop singing and start growling when I entered his territory, and in this case the nest was only half-built.

Eggs.

Four to five. Rather a late breeder as a rule, the 25th May is about the best average date. One early Spring, the nests I found in Sussex were all hatched on this date. But in the following year, which was very late, I knew of one bird that did not start to sit until 5th June.

If its nest is destroyed it will replace it and the eggs in from 10 to 12 days. Single brooded.

Nest finding.

When I was a boy I spent a lot of time in searching in thick jungles round where the cock was singing, looking for its nest. I had been told that these birds always nested in inaccessible places. Now I know better. The Nightingale sings in the thickets but nests in the open. The cock may sing a long way from the nest, which is often found by the roadside, or hard by a small track or ride through the woods. At the base of a bed of nettles, or even a small group of them, is a popular place. Tree stubs on banks, the base of nut-trees, against a telephone pole, in dog-mercury, butcher's broom, long grasses, and once in a gorse-bush are some of the places I have seen nests.

I always get a thrill when I look into one of these nests; they are so deep that one thinks they are empty until right over them.

They are not too hard to find by searching in the sort of positions I have enumerated if a start is made before the grasses get high.

The hen alone incubates and I have not myself seen her fed on the nest; but Carlyon-Britton tells me that he once found a nest by watching the cock go to his sitting mate with food.

THE RED-BACKED SHRIKE.
(Lanius c. collurio).

Field characteristics.

Male very striking, a little bigger than a Yellow-Hammer, head—grey, back—red, underparts—white, and broad black bar running through eye. The hen is drab, being dull chestnut, underparts dirty white barred with black.

When perched, Shrikes seem to hold their tails lower than most birds, and they can be identified at long range by this habit. The flight is Woodpecker-like, a series of dipping swoops, until finally they alight on the topmost twig of some bush or tree.

Allied species.

The Woodchat Shrike is the same size but the cock is black and white and hen brown and white.

Song.

An attractive warble, but the song is not often heard.

Call and alarm notes.

The alarm is a harsh ' chack.'

Eggs.

Four to six.

The 25th May is a good date on the South coast, but in this part of Hampshire laying does not

start until June is in. Shrikes replace a lost nest and eggs regularly in 10 days. Single brooded.

Nest finding.

I have heard it said that the nest of the Butcher bird is the easiest nest in the land to find. In fact all one has to do is to throw a stone at the cock and he will go straight to the nest! It is not quite as easy as that. However one cannot help but see the cock on entering his territory, for there he sits perched on the highest object in his domain. Every now and then he drops to the ground, captures a beetle and returns to the same post. Or he may fly straight to the nest with it and feed his mate. Then it is easy if you can follow his flight, but they are addicted to nesting in impenetrable jungles of thorns, brambles, and in Hampshire large gorse brakes. In such uninviting situations I have found that it pays a good dividend to concentrate on watching the place he reappears at, after feeding his mate. I have found that this is usually the spot where the nest is, and that it will be an appreciable distance from where it seemed to vanish with food in his beak.

D. W. Musselwhite and I were the first to record the fact that all nest-building is done by the cock Butcher-bird. The hen is seldom in evidence until the young are out of the nest. I have seen the cock build at all times of the day, even late in the evening. He also may be found sitting on the completed nest just before laying commences. Once only, in Germany, have I flushed him off eggs.

COMMON SNIPE
(Capella g. gallinago).

Field characteristics.

A small plump bird the size of a thrush with a short tail. Brown on top and whitish grey below.

On the ground its absurdly long bill at once distinguishes it from other birds except the Woodcock, which is much larger. When flushed it flashes away in a series of zig-zags uttering a harsh ' sha-arp.'

Allied species.

None in breeding season.

Song.

No true song, the well known ' drumming ' which is like the bleating of a goat, is made by the wind on the two extended outer tail feathers when the bird is in a dive.

Call and alarm notes.

A loud clear ' chuck-er ' repeated rather slowly as the bird races overhead.

Eggs.

Four, sometimes only three, while five are rare.

An early breeder in the South, many are sitting hard by the first week in April, or even earlier. Many early nests are destroyed by floods. Fresh eggs have been found in June.

Some may be double-brooded. They will replace a lost nest and eggs in 12 days.

Nest finding.

The favourite nesting site is rushy wet meadows in river valleys, the margins of lakes, ponds, sewage-farms, etc. Snipe betray their presence by their ' drumming,' which can be heard a long way off.

When many pairs are breeding, say in a large expanse of marsh, the best way to find nests is by walking up in line as in partridge shooting.

One day in the Arun valley with three friends we found in this way twelve nests.

Sitting Snipe have very keen hearing and are liable to leave the nest too far ahead for the place

NEST OF RED BACKED-SHRIKE.

Photo by Capt. J. H. McNeile.

CUCKOO'S EGG IN NEST OF PIED WAGTAIL BUILT
AGAINST SIDE OF HOUSE.

Photo by G. Tomkinson.

NEST OF NIGHTINGALE.

Photo by Rev. C. J. Pring.

NEST OF DARTFORD WARBLER.

Photo by Rev. C. J. Pring.

to be marked. Silence is essential, as nothing disturbs them quicker than the human voice. Therefore no talking must be the order, and all communications should be by signals. It is best to walk up-wind; it means retracing one's steps after each strip, to do so, but it is well worth it. It also pays to look well ahead, for more nests will be found by seeing the bird leave than ever there will be by ' spotting ' the eggs during the walk.

On certain days, Snipe sit so lightly that every bird seems to leave the marsh as soon as foot is set therein. All that can be done then is to try and watch them back, and mark the spot where they drop. But a big ' bag ' will not result from these conditions, and it is only by finding many nests that there is a chance of getting something really good.

THE STONECHAT.

(Saxicola torquata hibernans).

Field characteristics.

A striking little bird. The black cap, white collar, and pink breast of the cock makes identification easy. The hen resembles a dull brown Robin.

Allied species.

The male Whinchat lacks the black head and white collar, and the pale line through the eye is easily discernible. The hen is not unlike a hen Stonechat but has the light eye-stripe.

Song.

Very like a Hedge-Sparrow's.

Call and alarm notes.

The alarm is a loud ' tick-tick-tick ' like knocking two stones together.   e

Eggs.

Four to six. Double brooded. The 12th April is a good date for first clutches, and late May or early June for the second brood. Young birds have been noted by A. E. Burras being fed by their parents in October, which looks as if a third brood may be occasionally raised. Nest and eggs if destroyed will be replaced in from 10 to 12 days. The hen alone incubates and the cock does not feed her on the nest.

Nest finding.

When the hen is sitting her mate keeps guard, perched on the tip of some bush, tree or telegraph-pole, etc. He has one or two favourite stances on which he may be generally found. The common error must not be made of jumping to the conclusion that any of these perching places are near enough to the nest to help in its location by simply beating about in an endeavour to flush the hen. The odds are heavily against one, and time and energy will be unnecessarily wasted by attempting to flush her straight away. The cock does command a view of the nest from where he stands, but he gives no clue to the observer even by the direction in which he is facing. The best procedure is this: Sit down or stand still for ten minutes and watch the cock. The hen may come off and join him to feed, or he may call her off himself, in which case it should not be too difficult to withdraw to a commanding position and watch her back. Length of feeding time varies much, according to whether they are shy or not. In fine weather they will stay off longer than in wet. A hen may return to eggs in a few minutes, or stay off an hour even if the eggs are incubated. Some I have known were most difficult and refused to return even after the observer had withdrawn 200 yards or more, and appeared to object to being viewed through field-glasses. Others pay little heed, and

will go down within fifty paces of the observer. One hen in particular which was returning from feeding kept us waiting for over twenty minutes and still would not go down. Eventually we changed our positions—still no luck—so we sat down in the long heather to rest. Suddenly she flew straight towards us, low over the ground, and disappeared into the heather almost at our feet. I took six paces forward and she fluttered off her nest!

Now—if after watching the cock for some time, the hen has not put in an appearance, what can be done to help matters?

A slow walk towards him will make him fly to another perch, a follow up may lead one right away until he will suddenly return to where he started. Or he may on one's near approach show signs of alarm—' tick-tick '—even coming close and fluttering overhead. Now is the time to tap around, keeping a sharp lookout behind. It is easy to flush a Stonechat, yet not see exactly from whence she came. Then she must be watched back from a distance and the spot marked. If in the first sally he has led one on a ' wild goose chase ' the other points of the compass should be tried one by one. Near the edges of paths first, as most nests are so placed.

I have heard it stated that the Stonechat's nest is an easy one to find. I cannot agree. It is quite true that it is not difficult to find an odd nest here and there, but if the searcher is ' plotting ' all the breeding pairs on a given area, an immense amount of time, patience and endurance is needed.

THE TREE-PIPIT.
(Anthus t. trivialis).

Field characteristics.

Much like a Meadow-Pipit but slightly larger.

Allied species.

See under Woodlark.

Song.

The cock soars into the air and sings on the descent as he parachutes down with wings and tail spread wide; ' see-er-see-er-see-er-see-er-wee-wee-wee wee-weeee.

Call and alarm notes.

A loud ' chip-chip-chip ' without pause.

Eggs.

Four to six. Dates very variable. From about the 18th May as a rule, but in early years may be 10 days sooner.

They are mostly double-brooded in the South, and these may be sought from about mid-June onwards.

Will replace a lost nest and eggs in 10 days.

Nest finding.

In my experience a most difficult nest to find. The birds hate being observed, even at a distance and through field-glasses. The hen has a maddening habit of spotting the would-be observer, however well he may be concealed. She then takes up a position on some tall tree, and alarms incessantly until he removes himself.

Even when building, an easy time to find most nests, I have watched a hen Tree-Pipit pick up some grass and fly right out of sight.

How then come the wonderful series of these eggs to be seen in collections?

The easiest nests to find are on railway embankments, and in some parts of the country many nest on that site, and can be quickly found by walking the banks. But I want to discuss here the best method of discovering the many nests which

are not, from the bird-nester's point of view, so
conveniently situated. In Hampshire the heather
clad commons always have their quota of Tree-Pipits
each spring, never very numerous, but wide-spread.
Here, then, each pair must be watched separately,
as seldom indeed can one so postion oneself as to
overlook two breeding sites. To add to the difficulty,
one never knows, as with most birds, when the
time is right for eggs. The Tree-Pipit arrives on
its breeding ground early in April, and may start
nesting the first week in May, or not until late in
in that month. Another factor against one is the
lack of help from the cock. He sings all day,
except perhaps for a couple of hours in the afternoon, usually from one of two favourite perches. I
have heard it suggested that the nest is mid-way
between the two. I wish it were as easy as that,
but I have not found any evidence to support this
theory. He seems to take little interest in the nest
until the young are hatched. He certainly does go
off with the hen to feed but does not accompany her
back to the nest like the Meadow-Pipit. As a rule
the first intimation one gets that they have returned
from their feeding ground is the arrival of the cock
on one of his singing posts. The hen meanwhile has
reached her nest unseen. That is of course if the
onlooker is well enough hidden to escape her notice.
A useful tip if the ground is suitable is to watch
from inside a motor-car. I have tried this with
success with several birds including Tree-Pipits. By
far the best time for watching is the last two hours
of daylight. The birds then are not so wary, and
the hen leaves her nest to feed and drink at more
frequent intervals than at other times of the day.
The afternoon is useless. The hen Tree-Pipit's
normal approach to the nest, is to fly into a nearby
tree—a pine in this part of the country—stay there
for several minutes while she sees that the coast is
clear, and then to drop suddenly down on to the

nest. This means that the watcher must not relax for a moment if he is not to miss her. Heedless of the bites and stings of various flying pests, he must keep his eyes fixed on the area under observation.

Friends have told me that they have flushed Tree-Pipits off nests by walking about the area of the cock's singing posts. All I can say is that they have had better luck than I have, the odds are all against it. The hen sits very closely, and may let you walk right by her without leaving the nest. Also I think that the chief attraction in bird-nesting is to find the nest by ' science ' rather than by ' brute force.'

Now and then a nest can be found with surprising ease: In July, 1941, I visited a tiny chalk quarry in the Cotswolds. A Tree-Pipit was singing on a small bush at its edge. All around were growing crops, and I wondered where his nest could be. The track to the quarry was bordered by a low grass bank. I looked along the bank nearest to me, and there was the hen sitting on her nest.

THE WHINCHAT.

(Saxicola rubetra).

Field characteristics.

The cock has back—brown with black marks, throat and breast—reddish, underparts—white, eye stripe—whitish. The hen shows less black and is generally duller.

Allied species.

See under Stonechat.

Song.

A loud clear warble, at times snatchy and bubbly, some notes like a Redstart's. I rate him a

first class songster, and it is surprising that the Whinchat is not more widely recognised as such.

Call and alarm notes.

A loud ' u-tic u-tic ' much like a Stonechat.

Eggs.

Five to seven. The date to expect the first nests to be ready is the 25th May in the South and a few days later in Scotland.

A second brood is reared late in June or early July. They will replace a lost nest and eggs in from 10 to 12 days. The hen alone incubates and the cock does not feed her on the nest.

Nest finding.

For practical purposes all that has appeared under Stonechat could apply equally well to the Whinchat. But there are differences, even if very slight between the two species, and they are hard to put into words. Whinchats begin to arrive on their breeding grounds late in April in Scotland. Yet I have never found a bird incubating a full clutch before the 25th May. The place was swarming with them, and I found three nests in one small field, two in five minutes.

There is an extraordinary variety in the plumage of the cocks, not seen in Stonechats. Some are very handsome with strikingly marked black heads and cheeks. Others are just the opposite and can barely be distinguished from hens when seen alone. Some pairs I found to be just as cunning and elusive to watch as their relatives can be. Some equally confiding.

The favourite site of the nest in Scotland was under tussocks of coarse grass, close to a road or railway. One such nest I found very easily—the cock called off the sitting hen as I walked towards

him, and I saw her leave the nest—and I went
straight to it. It was on the lip of a little pit in
a meadow, in the sort of place one would look for a
Robin's nest. There was another nest with eggs
about fifty yards away, and this was under a tuft of
grass.

I think that on the whole the cock's singing
places are nearer to the nest than the Stonechat's,
and I am convinced that they are shyer builders
than the latter bird, or that building takes place
very early in the morning.

The Whinchat is very much more numerous,
especially in the North country and Wales. The
nest is not difficult to find if incubation has started.
The hardest problem is to know when to commence
the search, owing to the long time lag between
arrival and nesting. This is most unusual with small
migrants.

Like the Stonechat, the hen comes off eggs very
frequently, and may remain away feeding for some
time. She has the same long low flight back, with
the cock in attendance. But owing to the more
open terrain does not have the same opportunity for
putting a gorse bush between the watcher and
herself.

THE WILLOW TIT.

(Parus ater kleinschmidti).

Field characteristics.

A black-capped tit, whose cap is said to be dull,
it has two buff bars on the secondaries.

Allied species.

The Marsh Tit has a shiny cap and lacks the
buff wing-bars, but my own opinion is that both

these distinctions are useless and misleading as means of identification in the field.

Song.

Difficult to describe in words, loud and clear with Nightingale-like notes, to my ear unlike any other British song bird with which I am acquainted. There is another, and more often heard song, with something like the Redstart's about it, and this has the vigour of the Canary's outburst. It has long pauses between phases, and is unexpectedly powerful for so small a bird. I am indebted to Walpole-Bond for first telling me about this second song.

Call and alarm notes.

A harsh twangy ' chay-chay-chay ' with a yankee accent, a note of surprising depth and power for such a little bird.

Eggs.

Seven to thirteen. The latter number, as far as I am aware having been found only by myself.

The 7th to 12th May is the time most birds begin to sit. In an early season some will be hatched by then. Single brooded as far as is known.

D. W. Musselwhite, who has made a special study of this family, tells me that the time taken to replace a lost nest and eggs is 15 to 17 days. This includes the boring of a fresh hole. He has known of one case where the whole operation was completed in the amazingly short time of twelve days.

Nest finding.

As the Tree-Sparrow is to the House, the Marsh Warbler to the Reed, and the Dartford Warbler to the Whitethroat, so is the gentle Willow-Tit to its vulgar cousin the Marsh-Tit.

It is over fifty years since this bird was discovered in Britain, yet many good ornithologists

are ignorant on the subject of this chapter. I
cannot understand why. The Willow-Tit is wide-
spread but nowhere numerous. When I came to
live in Hampshire I had never seen the bird, but
was naturally keen to meet with it as soon as
possible. I picked it up almost at once, by hearing
its call note in the woods during the autumn. I
devoted the first twelve days of May in searching
for its nest, and was lucky to discover three in one
Wood.

The Willow-Tit is, I consider, a silent bird, and
must be sought for energetically. Many a time I
have watched a pair of black-capped tits feeding
in the birches above me, waiting for them to make
some sound whereby their identity would be
revealed. Once I was so engaged with the late
Robert Blockey, of the Haslemere Museum (a
good ornithologist and charming companion who was
so tragically killed over Leipzig with the R.A.F.).
He was certain that the birds we were watching were
Willow-Tits, and he said that he could see their
light wing-bars. After about half an hour one of
them gave vent to the well-known call of the Marsh-
Tit.!

The Willow-Tit must be looked for where there
is plenty of well rotted wood, so soft that one can
poke a finger into it. There is no evidence to suggest
that the ground must be wet or marshy. Some of
my nests have been found in very dry spots. I
believe that it is because there is naturally more
rotten timber in swamps that this tit is found there
and not because of any particular liking by the
birds for water. I have found nests in hedgerows
by the road-side, and in private gardens, but the
majority have been in the woods. So then a wood
with plenty of rotten trees should be sought (birch
is the best in this district, but alder is good too),
the area divided into sections, each walked in narrow

strips, and the rotten trees tapped with a stick. It is a characteristic of the Willow-Tit that she flushes easily from the nest in most cases, unless off young or highly incubated eggs. Only one Marsh-Tit have I known to do likewise, she had a nest at the top of a rotten and slender birch. I noticed the hole, tapped and off she came. Surely a Willow-Tit I thought, but her raucous cries soon contradicted me. Another curious fact I have noted is that in the great majority of nesting sites the hole faces SOUTH. As is well known, tits as a rule choose a North-Eastern aspect, and I am at a loss to understand why this bird should differ from the rest of the family in this respect. The holes are not very hard to see, and are betrayed by the saw-dust below. The eggs are not covered during laying as is normal with other tits. The nest holes are seldom higher than 18 feet, and some will be quite low. A proportion of the high ones will be inaccessible—too rotten to climb, or hold the weight of a ladder.

THE WILLOW-WARBLER.

(Phylloscopus t. trochilus).

Field characteristics.

A small greenish bird, paler underneath, which haunts the tops of trees and bushes, restless in its actions like a tit and seldom descending to the ground, except for drinking, until engaged in nest-building.

Allied species.

Differs from the Chiff-Chaff in having the legs LIGHT brown instead of black, and from the Wood-Warbler by its smaller size and duller colour.

Song.

A sweet rippling warble, suddenly rising only to die away.

Call and alarm notes.

A soft ' hoo-e,' uttered by both sexes.

Eggs.

Six to seven is usual in about equal proportion, but I have often found only four in a first nest, probably owing to the hen being old.

In the South the 12th May finds many Willow-Warblers sitting, but the 16th is a more reliable date and plenty can be found with fresh eggs a week later, but of course they may have lost their first nest.

This bird will replace a lost nest and eggs in 10 days. Often double-brooded in the South, when four to five eggs are laid early in July.

Nest finding.

The morning is a good time to watch for Willow-Warblers building, especially after a shower of rain. A position near a singing place of the cock should be taken up and a look out kept for the hen. Sometimes as she approaches the nest with material, the alarm will be uttered, and by the cock too, although he often pays no attention to the work but sings away as usual.

When the hen is sitting, by tapping round likely places (bracken, brambles, grass, etc.) in the neighbourhood of the cock's singing posts, she may be flushed from the nest. But so cunningly is it hidden, and so silently does she leave, that oft times she eludes the eye, and the plaintive alarm note is the first intimation that the ' bird has flown.'

Now a position should be found where she can be watched as she flits restlessly from twig to twig.

With luck she should be seen to go lower and lower until the final dive into the nest. In practice, owing to the foliage a hen often eludes one, the alarm ceases and the observer knows that she has slipped on to the nest unseen. I have known them to return to the nest without a sound. The time taken by the hen to return to eggs varies, from a few minutes to three quarters of an hour. If she refuses to go down and lingers about the same spot, the watcher is probably too close and must choose another place to stand or sit.

Willow-Warblers are partial to the edges of rides in woods and a number of nests are situated a bare two paces from the path.

Incubation is performed by the hen alone, and the cock does not feed her on the nest.

THE WOOD-LARK.

(Lullula a. arborea).

Field characteristics.

The white eye-stripe, short tail, and bat-like flight should leave no doubt as to this bird's identity. Its flight is more undulating than the Skylark's.

Allied species.

The Skylark is bigger with a longer tail.

The Tree-Pipit can be confused with it on the ground, but the Pipit has white outer tail feathers which the Wood-Lark lacks.

Song.

Very distinctive and carries over a long distance. Can be heard at any time of the year in mild weather. It sounds like ' whee-hoo-whee-hoo-whee-hoo-hoo-hoo-hoo-hoo-hoo.'

The notes are pure and flute-like, and once heard are never forgotten. After the Nightingale, I think this our finest songster.

Call and alarm notes.

A curlew-like ' whee-la-loo.'

Eggs.

Three to four, sometimes five in second broods. Best dates:

  1st brood — 30th March to 7th April.
  2nd brood — 12th May to 20th May.

The Wood-Lark will replace a lost nest and eggs in 10 days.

Nest finding.

Let us suppose that the reader has never met with this bird and is out to seek it for the first time.

How does he proceed?

In the first place the Wood-Lark likes quite a different sort of habitat from that of the Skylark, which is a bird of the meadow-lands, corn-fields, etc. The Wood-Lark loves the heath lands, pine trees, downland, in a word—poor land. A typical site is a barren field, the grass close-cropped by rabbits. Here and there are patches of bracken, a birch of two, and most noticeable of all—ragwort.

The area is walked in strips not wider than, say, twenty yards, the birds if there will probably be flushed, or seen on the ground. If so a glance through the glasses will disclose the white eye-stripe.

Then one should withdraw some distance, 200 yards if possible, sit down and watch. If they are near the nest, the hen will, sooner or later, walk to it and incubate, the cock will fly to one of his singing trees. If building the hen will take material to the nest. If they are feeding then

they may be expected to fly away to the nesting ground. This may be in the same field or out of sight. If the latter they must be followed up, not always as easy as it sounds.

If on the other hand, both birds are flushed instead of being seen on the ground, they will fly away and settle on the ground some 150 yards in in front, when they must be followed up. This procedure may be repeated several times; then instead of going ahead, both birds break back, uttering their flute-like call. The hen then should go directly to the nest, and the cock may go with her and settle on the ground nearby, or a bush or tree. If they are building or the clutch is incomplete, they will hang about near the nest, and the only thing to do if after watching, no building is seen, is to have a search. Early nests in March are not hard to see. Under a thin bramble is a popular site, or a tuft of dry bracken. Some are right in the open, even on places that have recently been burnt. Others are under tufts of heather, and all have a ' run ' up to them which catches the eye. I have found nests with a little canopy of grass over them, one such I presented to the Haslemere Museum.

Now let us suppose that in the search over likely ground no birds have been flushed, but suddenly the cock is heard singing high overhead. What then? I think it is best to walk slowly about, round likely spots, watching the singer meanwhile, rather than sitting down and waiting, for he may continue singing for a very long time (they usually do when waited for). If one happens to approach the sitting bird, the cock is likely to stop singing, fly overhead and sound the alarm. Once I was

taking a short cut across a bare fallow field, and a Wood-Lark was singing some way ahead, Suddenly I heard him give the alarm, high above me. I stopped dead in my tracks, looked around and there was the hen on her nest not a yard from my feet!

If the cock does not oblige by helping, there is a chance that the hen may be flushed, either from the nest, or from somewhere near it.

I always have a quick search round when I flush a Wood-Lark even some way ahead.

R. Carlyon-Britton asures me that it pays to walk backwards when the cock is present, as he calls off the hen when one has passed by. Then there is a chance to see from whence she came. This has not been my experience and I have found scores of nests. When a cock is on a tree, and shows a strong inclination to ' stay put,' then I think it is a sure sign that the nest is close by. The hen sits like wax, and her feathers harmonise well with the surroundings.

The Wood-Lark is prone to desert her nest either when under construction or when laying is in progress. Once the clutch is complete, I have never known one to forsake her nest.

The hen is not fed by her mate.

I once flushed a cock off four fresh eggs.

The chief difficulty in locating these birds on new ground is that the cock is such an uncertain singer. Many a time, on beautiful mild sunny days I have wandered over known breeding grounds in the height of the season, and have not heard a note. Conversely in a bitter North-East wind some have sung well. One year a pair nested six paces from my garden fence, yet I only heard the song on very few occasions.

THE WOOD WARBLER.

Phylloscopus sibilatrix).

Field characteristics.

A small greenish-yellow bird, underparts whitish, it has a yellow eye-stripe.

Allied species.

Differs from the Willow Warbler and Chiff-chaff by its larger size and brighter colouring ,although in its natural haunts (the tree-tops) these differences are by no means easy of recognition, but as soon as a sound is uttered its identity is assured.

Song.

Loud and pentrating, described as a shivering ' sip-sip-sip-sip—tr-r-r-r-r-r-eee.'

Uttered on a tree or on the wing as the bird flutters from one tree to another.

Call and alarm notes.

A loud and plaintive ' pee-pee ' repeated after an interval of two or three seconds, reminding one of the piping cry of the Bullfinch only louder and clearer. Those of the cock differ slightly from the hen, being louder, but are not always easy to distinguish.

Eggs.

Five to seven. Early clutches can be found on 21st May, but the 28th of that month is the best all round date, whilst some may not be completed before 6th June. A second brood is sometimes reared, and these nests usually contain but four eggs, late in June or early July.

Nest finding.

One of the easiest of the small birds' nests to find, I have inspected as many as thirty in a good

season, or thirteen in twenty-four hours, although
this bird is nothing like so numerous round my
home as it is in parts of Wales or the New Forest.

Once the peculiar and distinctive call note of
the hen is mastered the location of the nest is only
a matter of time. These birds allow a much closer
approach than Willow Warblers and Chiff-chaffs
provided the observer keeps reasonably still. Every
year a proportion of cocks fail to obtain mates, yet
these birds frequent the same spot all day, singing
and behaving as if the hen was sitting on its nest.
So much valuable time can be wasted by anyone
' down for the day ' watching for non-existent hens.
But if one is living in Wood-Warbler country and
passing their singing haunts frequently from time
of arrival (in late April) until mid-May without
hearing the tell-tale ' pee-pee,' then one auto-
matically registers that bird as mateless and no
time is lost looking for a nest which does not exist.
Once the cock has found a mate, little time is
wasted before building starts. The hen may be
observed inspecting likely places on banks, amongst
bracken, bilberries, etc., and I have noticed that
they show a marked partiality to the proximity of
holly trees. On and off she keeps up her monotonous
call all day. At last the site for the nest is chosen
and only four of five days are needed to complete it.
Wood Warblers well repay watching during courtship
and building, and if lucky one may see the charming
love-flight of the cock. With trailing legs he parachutes down almost to the nest, loudly proclaiming
the first phase of his song ' sip-sip-sip,' and
apparently quite oblivious of the watcher. If too
late for the building operations, one arrives on the
scene when the hen is sitting, what is now the
procedure? It is a good idea to have a quick walk
round tapping likely spots or even to clap the hands,
which may bring her off.

The cock sings a long way from the nest but joins his mate as soon as she comes off to feed. Then she must be kept in sight (not letting her get too far or she may be lost) until she drops lower and lower, calling all the time, and finally drops on to the ground. The hen if incubation has started seldom stays off for more than fifteen minutes, but cases occur when they will stay away for as long as three quarters of an hour.

If the clutch is incomplete she will visit it from time to time, but it will be at long intervals. However at these times she will give away the approximate position of the nest, when a quick look round will reveal it, or a return in a few days' time will prove satisfactory.

The Wood-Warbler leaves the nest once an hour to feed, but some hens seem to be more often off than on!

Incubation is performed by the hen alone, and she is not fed by the cock when on the nest. I have seen both birds enter the nest together when there were eggs, and how they managed to find room in such a small space without breaking the eggs is a mystery.

THE YELLOW WAGTAIL.

(Motacilla flava flavissima).

Field characteristics.

The smallest of our British Wagtails, with a shorter tail than either the Pied or Grey Wagtails. The cock has a brilliant canary breast and underparts, with back a dull greenish-yellow. The hen is altogether duller, but both have the distinctive yellow eye-stripe and chin.

Allied species.

In the Blue-headed Wagtail (Flava flava), which is the continental form of our bird, the cock has a pale blue forehead—crown and nape, with a distinctive white eye-stripe and chin. The hen is similar to the Yellow Wagtail but with eye-stripe and chin whitish instead of yellow.

The pale headed form resembling M.beema which occurs now and then on the Kent and Sussex coasts, and which has an almost white head, is believed to be just a variety of flava.

Song.

A soft sweet warble, not unlike the Grey Wagtail's.

Call and alarm notes.

The call note is ' sizick-sizick,' and it is softer than the Pied Wagtail's cry. The alarm note, which is frequently uttered on the wing is very distinctive, a soft yet penetrating ' swee-ou-swee-ou.' Walpole-Bond in ' A History of Sussex Birds ' describes it as ' see-ou.'

Eggs.

Four to seven. The best all round date for fresh clutches is 23rd May. A second brood is reared in July.

The Yellow Wagtail will replace a lost nest and eggs in 10 days.

Nest finding.

Once the cock has been located, this I think, like the Wood Warbler, is an easy nest to find, provided the eyesight is good. What usually happens is this—on approaching the nest-site the hen either slips off, or is called off by the cock, when they fly overhead calling ' swee-ou.' All one

has to do is to retire quickly 100 yards if possible—keep still and watch. In a surprisingly short time the hen will fly over the nest site, or very near, dip or hover with wagging tail—take fright—and away again. This routine may be repeated several times, but at last she will alight on a projecting object such as a nettle, dock, long stem of grass, etc., and after a good look round, dive into the herbage on to the nest.

Now in one district in Sussex at any rate, the fun has only just begun, for as soon as one moves forward she takes fright, leaves the nest, and it is necessary to begin all over again, unless the position has been fixed by some mark or other. But normally, once back on the nest the hen sits like a rock, and does not leave until one is right over her.

Most of the incubation is done by the hen. Walpole-Bond says that the cock incubates at times. The hen is not fed on the nest by the cock.

In conclusion, the author hopes in some future edition of this little book to deal with further species. In the meantime, if these few notes help to make country holidays more interesting and happy to the budding ornithologist, then the time and patience spent on this subject has been well worth while.

INDEX

INDEX

www.ingramcontent.com/pod-product-compliance
Lightning Source LLC
Chambersburg PA
CBHW030552270326
41927CB00008B/1622

Ornithology

Ornithology is a branch of zoology that concerns the study of birds. Etymologically, the word 'ornithology' derives from the ancient Greek ὄρνις *ornis* (bird) and λόγος *logos* (rationale or explanation). The science of ornithology has a long history and studies on birds have helped develop several key concepts in evolution, behaviour and ecology such as the definition of species, the process of speciation, instinct, learning, ecological niches and conservation. Whilst early ornithology was principally concerned with descriptions and distributions of species, ornithologists today seek answers to very specific questions, often using birds as models to test hypotheses or predictions based on theories. However, most modern biological theories apply across taxonomic groups, and consequently, the number of professional scientists who identify themselves as 'ornithologists' has declined. That this specific science has become part of the biological mainstream though, is in itself a testament to the field's importance.

Humans observed birds from the earliest times, and Stone Age drawings are among the oldest indications of an interest in birds, primarily due to their importance as a food source. One of the first key texts on ornithology was Aristotle's *Historia Animalium* (350 BC), in which he noted the habit of bird migration, moulting, egg laying and life span. He also propagated several, unfortunately false myths, such as the idea that swallows hibernated in winter. This idea became so well

established, that even as late as 1878, Elliott Coues (an American surgeon, historian and ornithologist) could list as many as 182 contemporary publications dealing with the hibernation of swallows. In the Seventeenth century, Francis Willughby (1635–1672) and John Ray (1627–1705) came up with the first major system of bird classification that was based on function and morphology rather than on form or behaviour, this was a major breakthrough in terms of scientific thought, and Willughby's *Ornithologiae libri tres* (1676), completed by John Ray is often thought to mark the beginning of methodical ornithology. It was not until the Victorian era though, with the emergence of the gun and the concept of natural history, that ornithology emerged as a specialized science. This specialization led to the formation in Britain of the British Ornithologists' Union in 1858, and the following year, its journal *The Ibis* was founded.

This sudden spurt in ornithology was also due in part to colonialism. The bird collectors of the Victorian era observed the variations in bird forms and habits across geographic regions, noting local specialization and variation in widespread species. The collections of museums and private collectors grew with contributions from various parts of the world. This spread of the science meant that many amateurs became interested in 'bird watching' – with real possibilities to contribute knowledge. As early as 1916, Julian Huxley wrote a two part article in the *Auk*, noting the tensions between amateurs and professionals and suggesting that the 'vast

army of bird-lovers and bird-watchers could begin providing the data scientists needed to address the fundamental problems of biology.' Organizations were started in many countries and these grew rapidly in membership, most notable among them being the Royal Society for the Protection of Birds (RSPB), founded in 1889 in Britain and the Audubon Society, founded in 1885 in the US.

Today, the science of ornithology is thriving, with many practical and economic applications such as the management of birds in food production (grainivorous birds, such as the Red billed Quelea are a major agricultural pest in parts of Africa), and the study of birds, as carriers of human diseases, such as Japanese Encephalitis, West Nile Virus, and H5N1. Of course, many species of birds have been driven to (or near) extinction by human activities, and hence ornithology has played an important part in conservation, utilising many location specific approaches. Critically endangered species such as the California Condor have been captured and bred in captivity, and it is hoped that many more birds can be saved in a like manner.

THE SCIENCE OF BIRDNESTING